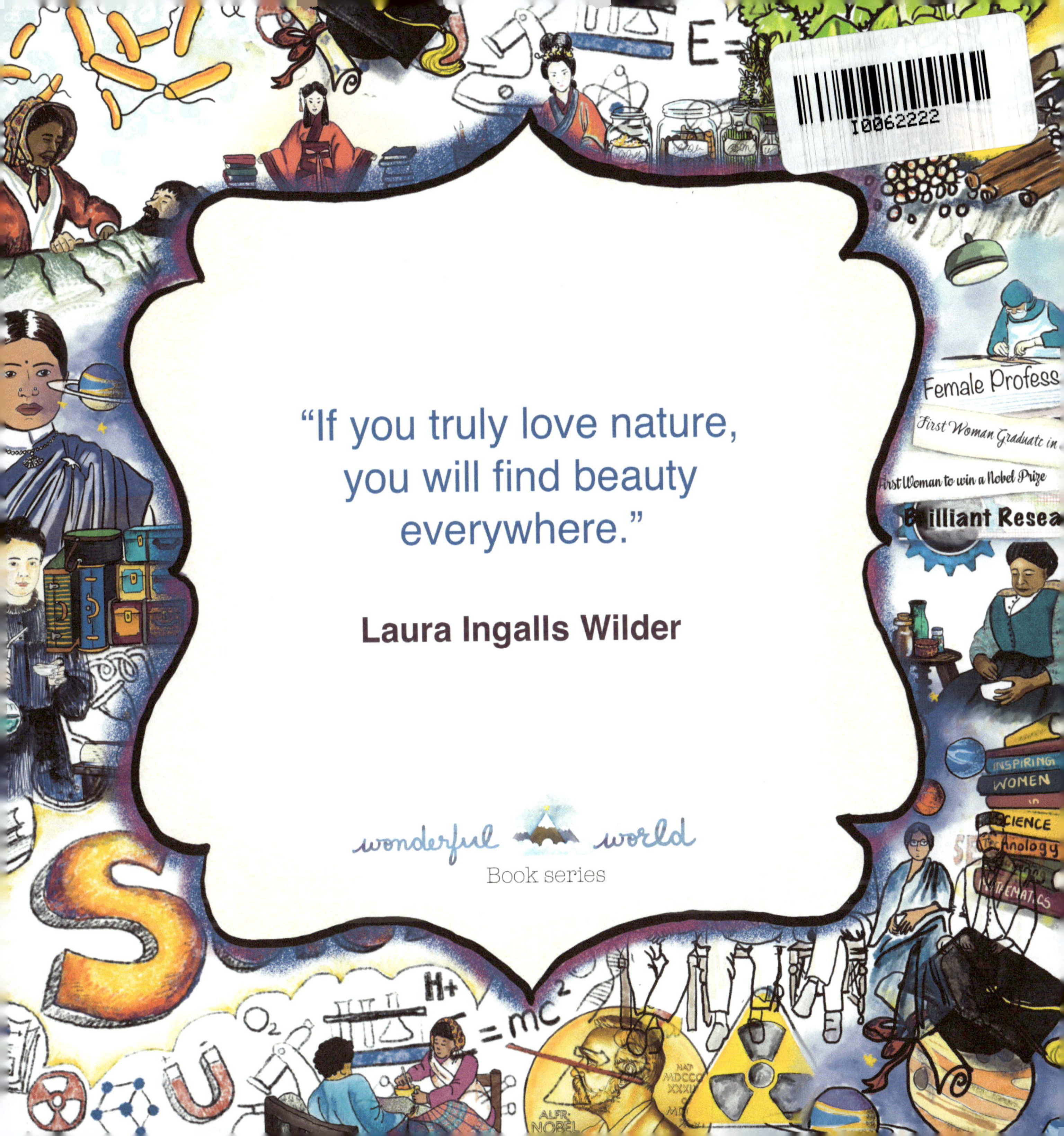
Female Profess
First Woman Graduate in
First Woman to win a Nobel Prize
Brilliant Resea
INSPIRING WOMEN in SCIENCE Technology MATHEMATICS
E = mc²
H+
O₂
ALFR. NOBEL

"If you truly love nature, you will find beauty everywhere."

Laura Ingalls Wilder

wonderful world
Book series

The shrubs and bushes and swaying trees,

The flowers and fruits and buzzing bees,

In her front garden,
and growing in the wild,

Fascinated Janaki (Jah-nuh-kee) even as a child.

Fascinated: Being really interested in something and wanting to learn more about it.

Born in India, in Thalaserry, (tha-la-sh-æ-ree)
She loved to study, and didn't want to marry.
So, Janaki made use of every chance,
To study more and more about plants.

Thalaserry: A city located in the state of Kerala, India.

She became a very famous botanist,

Botanist: A scientist who studies plants.

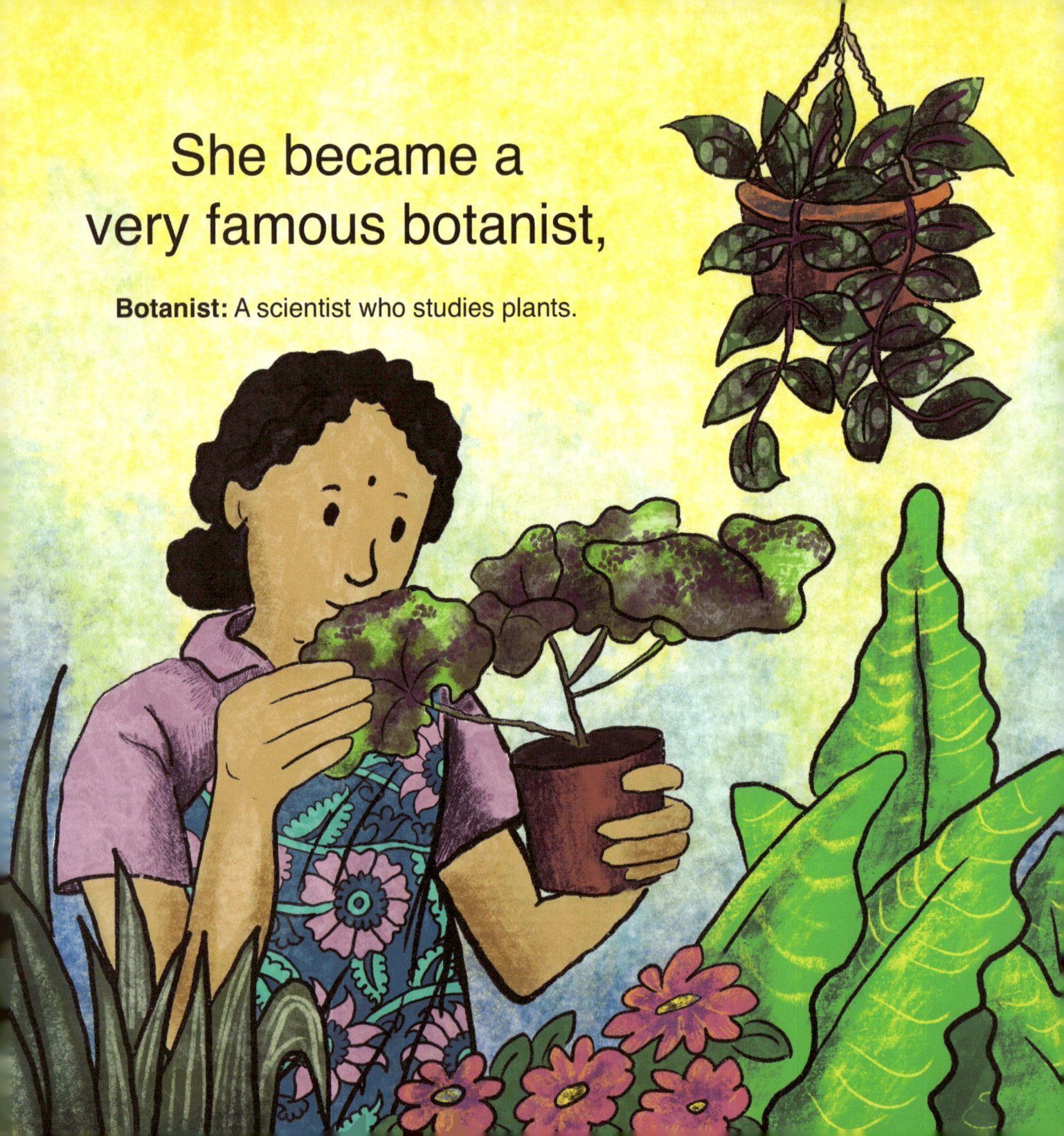

And a very dedicated cytologist.

Dedicated: Caring a lot about something and working hard to learn more about it.

Cytologist: A scientist who studies cells.
Cells are the building blocks which make up all living things.

Which means she studied plants and their cells

And knew what to do to make them grow well.

With her flowing hair and shining silk saree,
She was often mistaken for a Rajkumari;

Saree: A dress worn by women in some Asian countries like India and Sr Lanka. They tie the saree, which is a long piece of cloth around themselves in a special way so it looks like a very pretty dress.

Rajkumari: Indian word meaning 'princess'.

Michigan: A state in the United States of America. It's known as the Great Lakes state as its surrounded by Lake Superior, Lake Michigan, Lake Huron, and Lake Erie.

Great Britain: A country in Europe.

First, when she went to study in Michigan,
And then when she went to work in Great Britain.

During World War 2, when bombs fell overhead,

World War 2 (1939-1945):
A big battle (fight) between two groups of
countries — the Allies and the Axis.

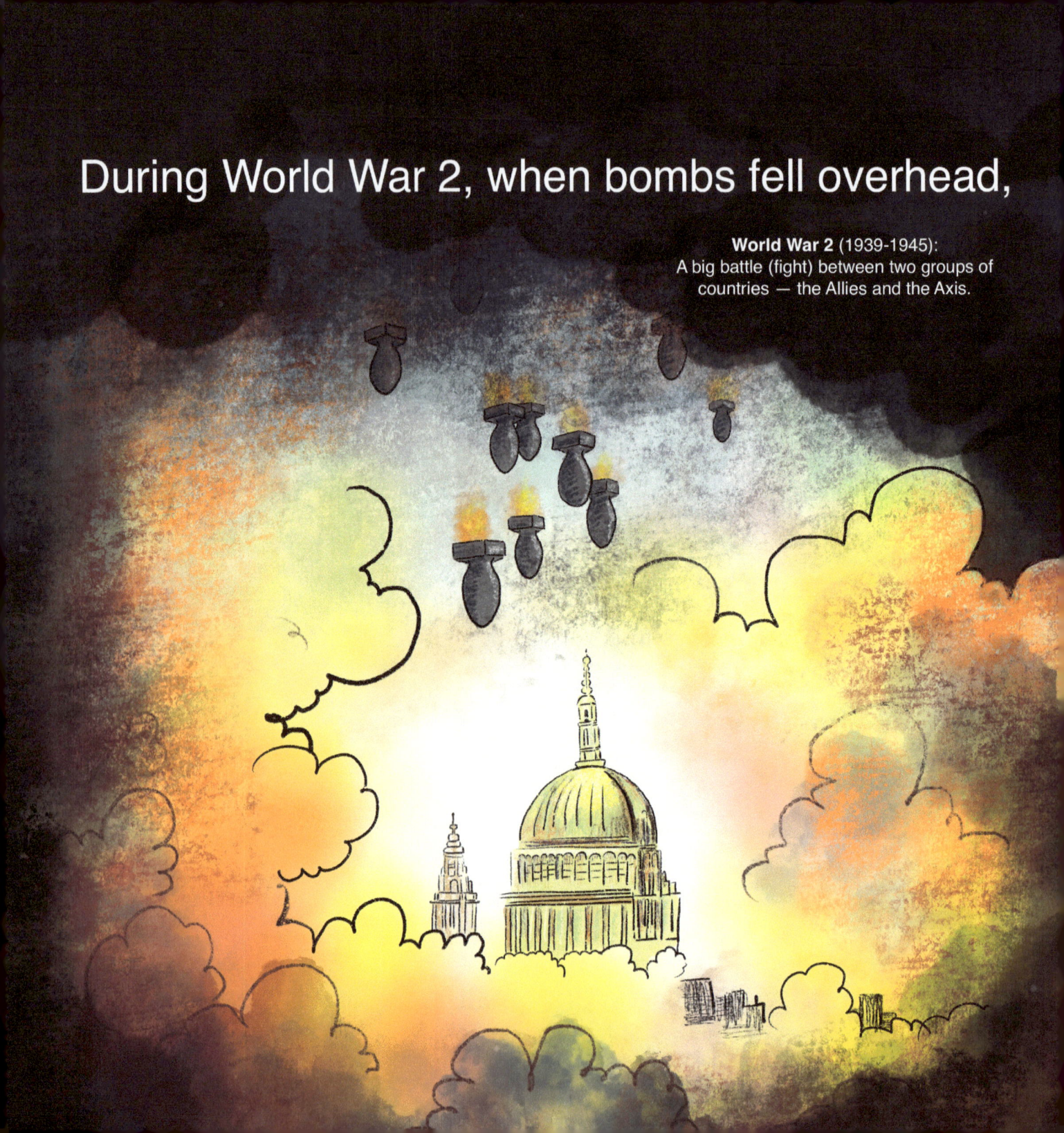

She'd dive under her table or beneath her bed,

Then clean up the broken bits of
glass and wood,

And resume her work as soon as she could.

Resume: Continue

She combined plants of different varieties, To create plants with unique qualities.

Varieties: Types

Qualities: The amazing things that make people, animals, or plants special. For instance, your qualities could be kindness, or the fact that you love ice cream etc.

That's how she made India's sugarcane sweet.

Sugarcane: A plant which is used to make sugar.

SUGAR

Saccharum officianarum
+
Saccharum spontaneum
= SWEETER, STRONGER
Sugarcane

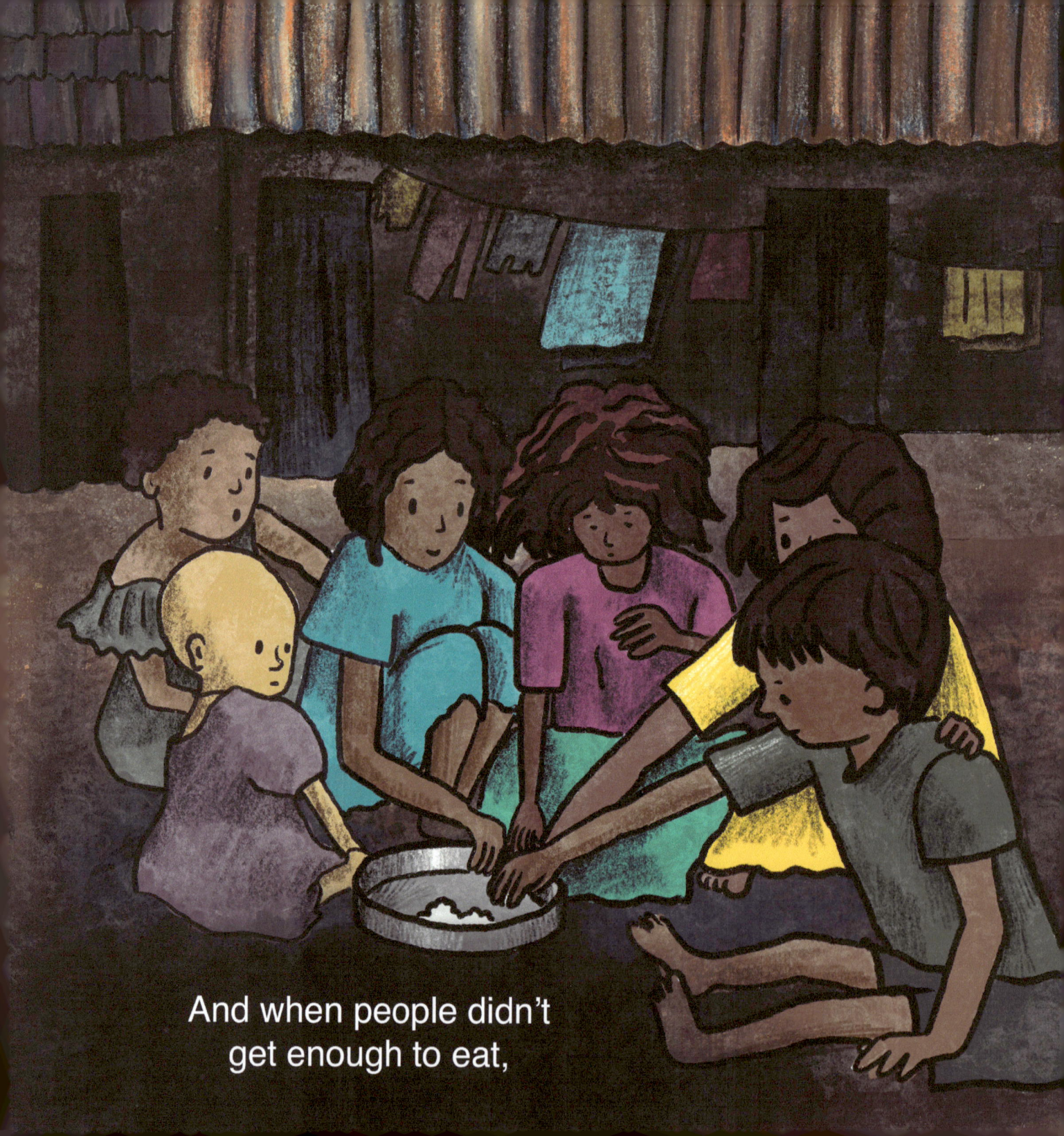

And when people didn't
get enough to eat,

Janaki made plants grow even faster,
to help fix this terrible disaster.

Disaster: When something bad happens.

She worked to preserve lush valleys,

Valley: A low-lying area between hills and mountains.

Forests, mountains, and all their trees.

And went
plant hunting
to find rare
breeds,

Breeds:
Types of living things

And grow
more of them
by planting
their seeds.

She knew the magic of our
life-giving Earth,
Of its potent power and
wonderful worth.

Like her, you can be
Earth's superhero too,

If you work to protect nature in all that you do.

Nature: Everything around us that isn't made by human beings. Plants, animals, the places they live in like forests, mountains etc.

Reduce waste, turn off the tap,

Reuse, or Recycle,

Buy less, plant more, walk,
or ride your cycle.

Just like the smallest seeds have tremendous powers,

And magically bloom into beautiful flowers,

Everything you do, no matter how small,
Helps to save the planet Earth for us all.

Janaki was
clever, **courageous**,
and **hardworking** too,

A wonderful person — just like you!

wonderful world
Book series

The Beginning

Thanks for reading my book.
I hope you've enjoyed it. For an independent author, ratings
are very important for the success of their book. I'd be
grateful if you could take a minute to rate this book on
Amazon/ Goodreads.
Your support makes all the difference.

Activity Pages
Timeline
Glossary

Cut Line

Fold Line

A simplified version
of the kind of house
Janaki Ammal
would have grown
up in

Colouring Page

Glossary

Botanist: A scientist who studies plants.

Breeds: Types of living things.

Cell: Cells are the building blocks which make up all living things.

Cytologist: A scientist who studies cells.

Dedicated: Caring a lot about something and working hard to learn more about it.

Disaster: When something bad happens.

Fascinated: Being really interested in something and wanting to learn more about it.

Great Britain: A country in Europe.

Michigan: A state in the United States of America. It's known as the Great Lakes state as its surrounded by Lake Superior, Lake Michigan, Lake Huron, and Lake Erie.

Nature: Everything around us that isn't made by human beings. Plants, animals, the places they live in like forests, mountains etc.

Glossary

Qualities: The amazing things that make people, animals, or plants special. For instance, your qualities could be kindness, that you like craft etc.

Rajkumari: Indian word meaning 'princess'.

Resume: Continue

Saree: A dress worn by women in some Asian countries like India and Sr Lanka. They tie the saree, which is a long piece of cloth around themselves in a special way so it looks like a very pretty dress.

Sugarcane: A plant which is used to make sugar. something bad.

Thalaserry: A city located in the state of Kerala, India.

Valley: A low-lying area between hills and mountains.

Varieties: Types

World War II: A big battle (fight) between two groups of countries — the Allies and the Axis.

Timeline of Janaki Ammal's Life

4th November 1897: Janaki Ammal was born to Diwan Bahadur Edalavath Kakkar Krishnan and Devi Kuruvayi. She was the tenth of nineteen children!

An avid student, Janaki Ammal studied in Sacred Heart Convent, Thalaserry, before going on to study in Queen Mary's College, Madras, and Presidency College, Madras. Needless to say, she got Honours in Botany!

1920 to 1923: She taught Botany at the Women's Christian College, Madras. All her sisters had marriages arranged for them, but when it was her turn, she refused, opting to continue studying.

1924: She won the Barbour scholarship to study at the University of Michigan. Despite this, she was detained at the Ellis Island, and was released because she was mistaken for an Indian princess!

1931: Janaki Ammal was awarded a PhD by the University of Michigan and became Dr E. K. Janaki Ammal.

1932-1934: Janaki Ammal was a Professor of Botany at the Maharaja College of Science, Trivandrum.

1935-1939: She worked at the Sugarcane Breeding Institute in Coimbatore. She cross-bred sugarcane to create sweeter varieties that grew well in India.

1939-1945: Janaki stayed in the UK during World War 2. She published the 'Chromosome Atlas of Cultivated Plants' along with C. D. Darlington.

1945-1951: She was the first woman scientist to be employed at the RHS Wisley. She studied the uses of a chemical called 'colchicine' which can help plants grow quicker and bigger.

1951: The Prime Minister requested her help in organising the Botanical Survey of India so Janaki Ammal returned to India.

1960s: She worked to preserve biodiversity in India's forests, and protected India's Silent Valley, home to many exotic animals and plants.

Until her death in 1984, she continued working as an Emeritus Scientist (like a Super-Scientist) in Madras, and did much to save plants from extinction.

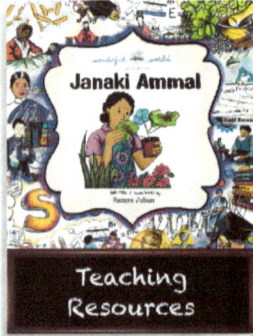

Teaching Resources

Check out the teaching resources at www.ramyajulian.com

A phenomenal botanist.and an eco-feminist before the term was invented, Janaki Ammal's story is very inspiring. There's little information about her, but every bit I read was fascinating.

Most educated women in Pre-Independent India tended to belong to the dominant castes, so I was pleasantly surprised to see that Janaki Ammal belonged to one of the suppressed castes. No doubt, she faced a lot of unpleasantness because of it. Often the only woman in conferences and meetings, she held her own and never gave up, though she lived at a time when a man's off-the-cuff remark over a cup of tea was given more weight than her most well-researched arguments.

Liberated, Passionate, Brilliant and a Breaker of Glass Ceilings, I hope you enjoyed reading about this unlikely hero.

My best,
Ramya

About the Author

Author, illustrator, and dentist, **Ramya Julian** finished her first novel at the age of ten and she avers it was very well received though it was read only by her brother.

She has all the hobbies of a maiden Victorian aunt – reading, writing, painting, crocheting, knitting and sewing, and the temperament of one. When she's not guilt-tripping her two daughters into good behaviour, she can be found devouring books, crafting poems and puns, and chuckling at her own witticisms. She grew up in India and now lives with her husband and their two daughters in London.

She has experienced so much joy through the enchanting artistry of many authors and creators, that she aspires to share at least some of it through her writing.

To see more of her work, visit **www.ramyajulian.com**

www.ramyajulian.com

Also in this series

NEXT IN LINE: MANY MANY MORE
WONDERFUL DIVERSE HEROES

SUBSCRIBE

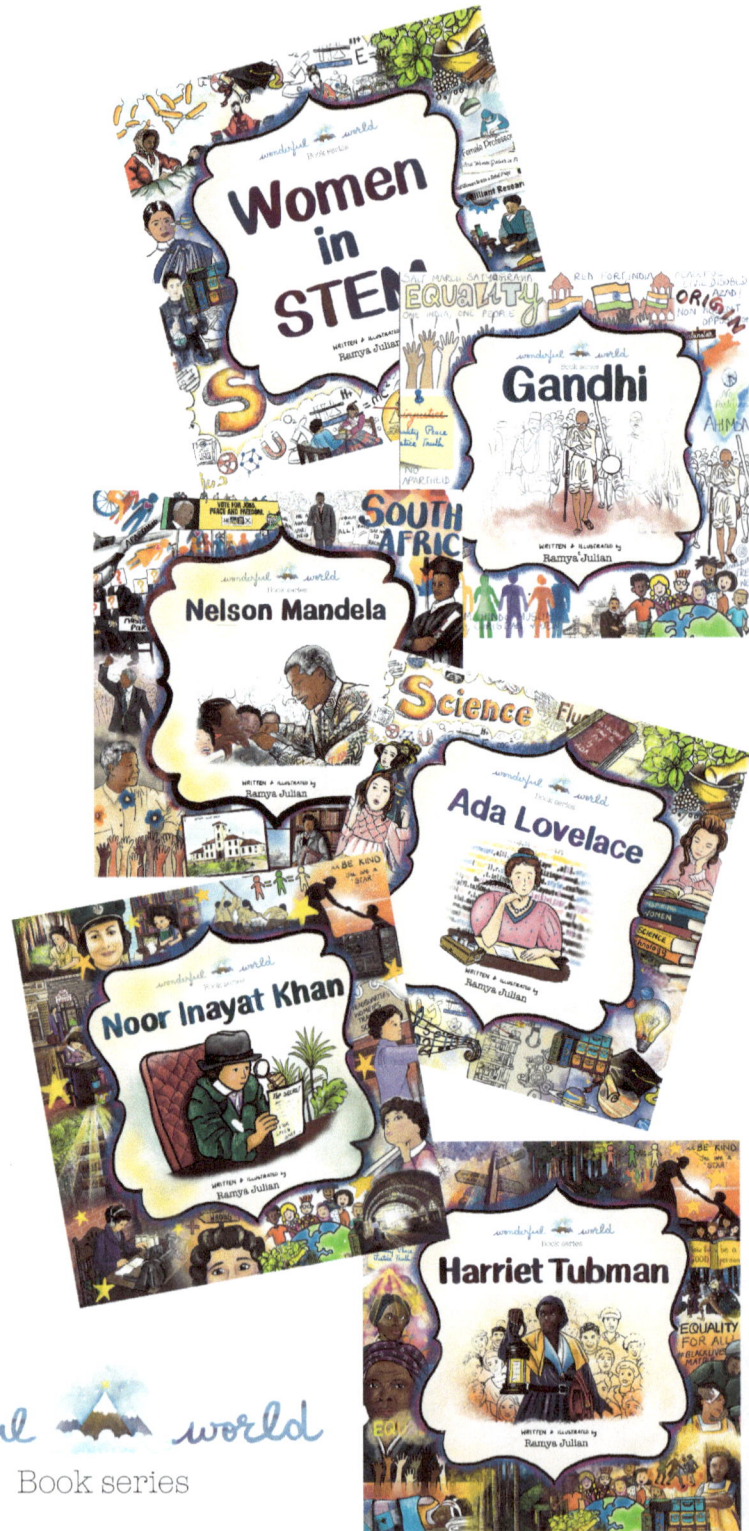

@RAMYAJULIAN

Women in STEM

Gandhi

Nelson Mandela

Ada Lovelace

Noor Inayat Khan

Harriet Tubman

wonderful *world*
Book series

www.ingramcontent.com/pod-product-compliance
Lightning Source LLC
Chambersburg PA
CBHW050912210326
41597CB00002B/96